ESSENTIAL GUIDE

MW00681921

the digital revolution

JACK CHALLONER

SERIES EDITOR JOHN GRIBBIN

London, New York, Munich,
Melbourne, and Delhi

senior editor Peter Frances
senior art editor Vanessa Hamilton
US editor Cheryl Ehrlich
DTP designer Rajen Shah
picture researcher Sarah Duncan
illustrator Richard Tibbitts

category publisher Jonathan Metcalf
managing art editor Phil Ormerod

Produced for Dorling Kindersley Limited by
Design Revolution Limited, Queens Park Villa,
30 West Drive, Brighton, East Sussex BN2 2GE
editors Ian Whitelaw, John Watson
designer Simon Avery

First American Edition, 2002

02 03 04 05 10 9 8 7 6 5 4 3 2 1

Published in the United States by
DK Publishing, Inc.
95 Madison Avenue
New York, NY 10016

Copyright © 2002 Dorling Kindersley Limited
Text copyright © 2002 Dorling Kindersley Limited

All rights reserved under International and Pan-American
Copyright Conventions. No part of this publication may be
reproduced, stored in a retrieval system, or transmitted in any
form or by any means, electronic, mechanical, photocopying,
recording, or otherwise, without the prior written
permission of the copyright owner.
Published in Great Britain by Dorling Kindersley Limited.

A Cataloging in Publication record is available from
the Library of Congress.

ISBN 0-7894-8417-X

Color reproduction by Mullis Morgan, UK
Printed in Italy by Graphicom

see our complete product line at
www.dk.com

contents

what is digital technology?

Y ou have probably heard the term "digital revolution" and almost certainly you have bought or used digital devices. If you have ever listened to music on a compact disc (CD) or an MP3 player, used a mobile phone, watched a film played on a digital versatile disc (DVD) player, or used a personal computer, then you have experienced digital technology. Also, the organizations that play important roles in developed countries rely heavily on digital technology. Banks, industry, communications, governments, and law-enforcement agencies, for example, all increasingly use digital technologies to create, communicate, organize, and entertain. The speed and ease with which digital technologies can process and transfer digital information is impressive, and has brought with it many benefits, but it also brings concerns over privacy and security. Understanding the fundamentals of how digital technology works will be increasingly important if we are to protect our freedoms, and get the most out of this exciting digital age.

at your fingertips
Even the humble computer keyboard is a digital device. It contains a series of switches connected to a microprocessor, which monitors the state of each switch and responds by producing digital representations of letters, numbers, other characters, and functions.

a digital world

Digital technology plays an ever-increasing role in the lives of people across the world – particularly those living in developed countries. Nearly all radio and television stations now record and edit their programs digitally, and some of them broadcast their output using digital signals. In many countries, it is uncommon to find anyone who doesn't own at least one device that makes use of digital technology – perhaps a CD or DVD player, a computer game console, or a mobile telephone; perhaps a digital personal organizer, a digital camera, or a digital video camera. And, of course, hundreds of millions of people regularly use computers to experience multimedia (sound, pictures, text, and video), much of which is accessed through the World Wide Web.

digital vision
The ease and flexibility with which digitized data can be transferred means that, for example, images captured by a digital video camera can be made immediately and widely available through other digital devices.

inside story

Despite our increasing familiarity with – and perhaps reliance on – digital devices, very few people have much idea what goes on inside them, or even what it means for a device to be classed as digital. Most digital devices are extremely complicated – only a highly qualified and specialized electronics engineer could hope to build a digital camera from scratch, for example. However, the principles on which digital devices are based are essentially simple, and understanding them gives an insight into the impact that digital technology is having on our way of life.

There are, of course, alternatives to most digital technologies. For example, typewriters can produce pages of neat text, just as computer printers can; photocopiers can reproduce drawings or photographs, just as digital scanners can; vinyl records and magnetic tapes can hold recorded sound, just as CDs or MiniDiscs can; old style, predigital telephone systems can transmit the sound of your voice to someone using a telephone anywhere in the world, just as the newer digital systems can. What unites all digital devices is that whatever is being copied, stored, or transmitted – sound, text, or images, for example – is represented as groups or streams of numbers (digits). As we shall see, this shared method of representing information has important consequences.

analog device
A vinyl record has a wavy groove that is a direct copy, or "analog," of the sound waves of the recorded sound.

why go digital?

In most cases, digital technologies offer significant advantages over the alternatives, and are rapidly becoming dominant. Since the 1980s, compact discs have replaced vinyl records and cassette tapes as the main medium in which people buy and listen to music. This is largely because they are more convenient (you can skip to different tracks at the touch of a button), more hard-wearing, and offer better quality. Handling sound digitally also means that endless copies can be made

digital device
A MiniDisc player does not carry direct copies of sounds. Instead, the sound is coded as numbers. Digital devices can carry sound, text, or pictures – or often a combination of all three – encoded as groups of numbers.

without loss of quality, and that it can be played by a host of different devices. This compatibility means that digital sound can be manipulated easily by computers and even delivered over the internet.

> **"Any significantly advanced technology is indistinguishable from magic. "**
>
> Arthur C. Clarke (1962)

There are many other benefits to "going digital": for example, digital telephone services offer faster and more reliable connections, along with a huge increase in the capacity of a telephone network; and digital television offers the possibility of watching programs on demand; digital networks (including the internet) offer the chance to communicate and share information quickly and easily.

the whole picture

Digital technology also brings potential problems. There are threats to privacy: digital systems can gather and communicate large amounts of information about people, often without their knowledge. There are environmental issues: devices can become obsolete quickly, requiring huge investments of energy and materials to replace them. There are also socio-economic issues: richer people and richer societies will always have greater access to the new technologies on which we may all one day depend.

In 1958, American electronics engineer **Jack Kilby** (b.1923) made the first integrated circuit – by creating a circuit containing several transistors and other components from a single piece of silicon. Not all integrated circuits are digital circuits, but those that are provide the computational power needed by digital devices in a very small space. In fact, the integrated circuit, more than any other invention, has made the digital revolution possible.

the world in numbers

At the heart of every digital device, you will find one or more microprocessors. These tiny electronic circuits are extremely complicated, but in essence they can only do one thing: process simple numerical calculations very quickly. Before sounds, pictures, or text can be stored and manipulated on a digital device, they must first be represented by numbers – a process called digitization. So, for example, images are digitized by a scanner or a digital camera, and a digital sound recording device can digitize sound.

transistors

In a microprocessor, the numbers representing text, sound, and images are themselves represented by electric currents passing through components called transistors (see panel on page 10). Typically, millions of transistors are etched onto a microprocessor, and each one is a tiny switch that can only be in one of two states: on or off. This corresponds to the counting system used in digital devices, which is based on the number two: the binary system. The number system we humans use – called denary, or base 10 – is based on the number ten and has ten digits – 0 to 9. The binary system has two digits: 0 and 1. A transistor represents 0 when it is open (nonconducting), and 1 when it is closed (conducting electricity).

doing the calculations
A microprocessor is an integrated circuit that processes digital information – simply by storing and adding numbers encoded as electronic pulses. A microprocessor inside a scanner processes numbers that represent the scanned pictures or objects, as well as communicating with a computer to which the scanner is connected.

scanner

microprocessor

how a transistor works

A positive electrical charge is sent down an aluminum wire that runs into the transistor, at what is known as the gate, and flows into a layer of conductive polysilicon surrounded by nonconductive silicon dioxide. The positive charge attracts negatively charged electrons out of the base of P-type (positive) silicon that separates two layers of N-type (negative) silicon. The electrons leaving the P-type silicon create an electronic vacuum, which is filled by electrons from a second terminal called the source. As well as filling the vacuum, the electrons from the source also flow to a third terminal called the drain. This completes the circuit, and turns the transistor on so that it represents a binary 1. If a negative charge is received at the gate, electrons from the source are repelled and the transistor is turned off to represent a binary 0.

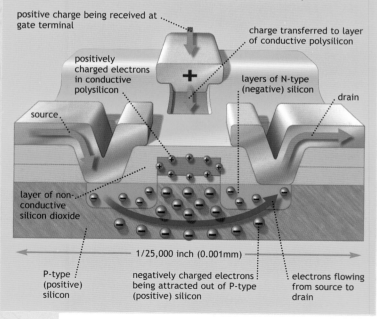

positive charge being received at gate terminal

charge transferred to layer of conductive polysilicon

positively charged electrons in conductive polysilicon

layers of N-type (negative) silicon

drain

source

layer of non-conductive silicon dioxide

1/25,000 inch (0.001mm)

P-type (positive) silicon

negatively charged electrons being attracted out of P-type (positive) silicon

electrons flowing from source to drain

The German mathematician and philosopher **Gottfried Leibniz** (1646–1716) originated the binary number system. Inspired by the "yin and yang" duality of the *I-Ching*, he worked out denary numbers as strings of 0s and 1s. He also pioneered the use of symbolic logic in mathematics. All modern digital devices use both binary numbers and mathematical logic.

binary numbers

The binary number system – also called base 2 – is founded on the same rules as the system known as base 10. Using base 10, any number can be expressed by using combinations of the 10 digits. The value assigned to a particular digit depends on its position in a number: in the number 333, for example, the right-hand three has a value of 3 (3 x 1), the next has a value of 30 (3 x 10), and the third has a value of 300 (3 x 10 x 10, or 3 x 10 squared, usually written as 10^2).

You can also write down any number using combinations of the two digits of the binary system. Again, they have different values depending on their place in a number. However, binary numbers look very different from the same numbers written in denary. For example, the number fifteen in denary is, of course, 15. In binary, the same number is written 1111, pronounced "one-one-one-one": from the right, the 1s have values of 1, 2, 4 (2 x 2 or 2^2, known as 2 squared), and 8 (2 x 2 x 2 or 2^3, known as 2 cubed).

binary pattern
Looking at the binary numbers from zero to fifteen, a definite pattern becomes apparent. The rules are the same as in denary (base 10), but there are only two digits available.

	2^3 (8)	2^2 (4)	2^1 (2)	2^0 (1)
0	0	0	0	0
1	0	0	0	1
2	0	0	1	0
3	0	0	1	1
4	0	1	0	0
5	0	1	0	1
6	0	1	1	0
7	0	1	1	1
8	1	0	0	0
9	1	0	0	1
10	1	0	1	0
11	1	0	1	1
12	1	1	0	0
13	1	1	0	1
14	1	1	1	0
15	1	1	1	1

representing binary digits

The transistors etched onto microprocessors are not the only means by which digital devices represent binary information. For example, streams of binary digits can be coded into radio waves, as sudden discontinuities (phase changes) or sudden changes in strength

how a cd stores information

The pits in the aluminum layer in a CD are the key to how a CD stores digital information: specific patterns of long and short pits represent specific groups of binary digits. The pits are arranged along a spiral track more than two-and-a-half miles (four kilometers) long. A CD can carry any digital information, such as text, pictures, video, and sound.

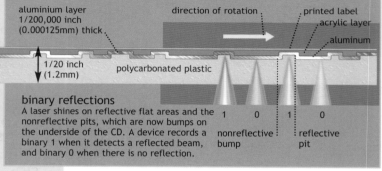

aluminium layer
1/200,000 inch
(0.000125mm) thick

direction of rotation

printed label
acrylic layer
aluminum

1/20 inch
(1.2mm)

polycarbonated plastic

binary reflections
A laser shines on reflective flat areas and the nonreflective pits, which are now bumps on the underside of the CD. A device records a binary 1 when it detects a reflected beam, and binary 0 when there is no reflection.

1   0   1   0

nonreflective bump

reflective pit

(amplitude). Binary digits can also be represented by tiny electric charges, which can indicate stored (1) or not stored (0), in the memory of a computer. Particles on the surface of a hard disk can be magnetized in one of two directions (0 and 1); the long and short pits in a CD also store binary 0s and 1s (see panel above). The many methods of storing binary digits means that digital information can be transferred easily between devices.

Storing, transferring, and manipulating information as binary numbers is convenient and dependable. Recorded sound on the hard disk of a computer can be played again and again without any loss of quality. A CD writer can make countless copies of music CDs, each one as good as the original; and, because text, sound, and pictures are digitized in standard formats, people can easily exchange digital information, typically via the internet.

bits and bytes

The digits 0 and 1 that are used to write down binary numbers are often called bits – short for binary digits. However, bits are often bundled into groups of eight, each called a byte – the basic unit of information held inside most digital devices. Distinct groups of bytes held inside computers are called files: a particular file may be a program, consisting of commands that the computer carries out, or a "document," which contains the digitized text, sound, or pictures.

bigger than bytes

One byte – eight bits – holds only a very small amount of information, and so the size of files is generally much larger than a few bytes, and so is the size of most storage devices. For this reason, most people are more familiar with the terms kilobyte (KB), megabyte (MB), and gigabyte (GB) than with bytes. It would be reasonable to assume that a kilobyte is a thousand bytes – just as a kilometer is a thousand meters. However, because the binary system is based on the number two, a kilobyte is actually 1,024 bytes ($1,024 = 2 \times 2 \times 2 \times 2 \times 2 \times 2 \times 2 \times 2 \times 2 \times 2$, or 2^{10}). One megabyte is 1,024 kilobytes – 1,048,576 bytes (2^{20}) – and one gigabyte is 1,024 times as large again: 1,073,741,824 bytes (2^{30}).

key points

• Digital devices represent text, sound, images, and video as binary numbers.
• Binary numbers are represented in digital devices in a range of ways.
• All digital devices have at least one microprocessor.

byte-sized
If a single grain of rice represents a bit of information, a byte consists of eight grains, a megabyte is represented by a sackful of rice, and a gigabyte is 1,024 sacks of rice.

1 bit   1 byte   1 kilobyte   1 megabyte

1 gigabyte

digital text

Many digital devices can store and manipulate text. Documents can be typed into computers, stored as files, and then edited and printed out; mobile telephones can send and receive short text messages; some digital sound devices allow you to enter the names of music tracks; DVD players display interactive text-based menus that allow you to select different scenes from a film; and electronic address books can be used to store hundreds of names, addresses, and telephone numbers.

coded messages

Inside digital devices, text is represented by using codes, in which a different collection of individual bits is used to represent characters, such as the letters of the alphabet. One of the standard codes, called the American Standard Code for Information Interchange (ASCII, pronounced "askey"), uses seven bits for each letter, giving a total of 128 different combinations. The first 32 of these combinations (from 0000000 to 0011111) are reserved for instruction codes, but the remaining 96 represent the letters of the alphabet (in lower and upper case), punctuation marks, mathematical symbols, the digits from 0 to 9, and the space between words. Most mobile telephone text messages, as well as paging and voicemail systems that are sent via the Short Message Service (SMS), use a code that is very similar and is called the 7-bit Default Alphabet.

Most computers actually use a code called Extended ASCII, which uses eight bits to represent each character. This has

say it in semaphore
There are many ways of representing words and numbers in coded form. In semaphore signalling, for example, each letter of the alphabet is represented by the positions of two handheld flags.

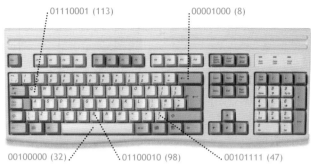

01110001 (113)   00001000 (8)

00100000 (32)   01100010 (98)   00101111 (47)

keying in
Pressing the keys of a computer keyboard produces streams of electric pulses, which represent eight-digit binary numbers. The pulses pass along a cable to an integrated circuit inside the computer's case.

the effect of doubling the number of combinations of bits to 256, making it possible to include more unusual and less-used symbols, and versions of alphabetic characters with accents for use in languages other than English. So in a computer, "letter space" is represented by 00100000 (ASCII 32), "b" by 01100010 (ASCII 98), and "/" by 00101111 (ASCII 47). The fact that the numbers of the Extended ASCII code are all eight bits long is, in fact, the reason why the byte is the basic unit of information in many digital devices.

text messaging

Pressing key combinations on a mobile phone generates seven-digit binary numbers. Each number represents a different letter, a number, or a punctuation mark. The seven-digit binary numbers are stored in the phone's memory, and are sent as text messages encoded into radio signals when the "send" command is issued.

R U COMIN OUT
L8R 2NITE ?

digital sound

Sound can also be represented digitally – as computer files or as streams or groups of bits on other digital devices. In addition to computers, technologies that make use of digital sound include CD players, MP3 players, MiniDisc players, digital radio and television, digital mobile telephones, and DVD players. Some digital devices are able to analyze the digitized sound of a person's voice, and recognize words. This speech-recognition technology allows people to control devices by using their voice, or even to dictate long passages to them. Before this can be achieved, sound must first be represented in

what is sound?

Sound is produced by vibrating objects, such as the vocal cords in your throat or the prongs of a tuning fork. When you strike a tuning fork, the prongs vibrate hundreds or thousands of times every second. The vibrations disturb the air, sending out waves of varying air pressure: sound waves. Sound waves are invisible, but we can visualize them as wavy lines called waveforms, which are essentially graphs of the changing air pressure. Each sound produces a different pattern of vibrations, and therefore has a different waveform.

sound waveforms representing changes in air pressure caused by evolution of sound waves digital form. Any device that records sound must be able to produce a representation of a sound's waveform. An important component of most recording systems is a microphone. Sound waves cause a diaphragm inside the microphone to vibrate, and this produces a continuous and varying electric current, whose variations match those of the air pressure in the original sound. This electric current is called an audio signal. Analog recording makes a direct and continuous copy, or analog, of the signal, and therefore of the pressure variations of the sound wave.

from waves to binary digits
Sound waves cause a diaphragm in a microphone to vibrate in response to air pressure changes. Those vibrations are converted into analog signals, which are passed to an analog-to-digital converter (ADC). Here, the signals are sampled to convert them into binary data.

sampling

Digital recording, however, produces a stream of binary numbers, rather than a continuously varying signal. The digitization of sound is achieved by an electronic circuit

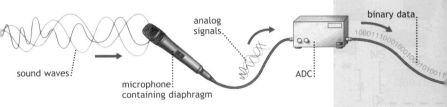

binary data

analog signals

sound waves

microphone containing diaphragm

ADC

called an analog-to-digital converter (ADC), which measures the strength of the audio signal thousands of times every second – a process called sampling. You can think of each measurement, or sample, as denoting the height of the waveform at a particular instant. Together, the thousands of samples – in the form of binary numbers – describe the variations of the audio signal, and therefore represent the shape of the waveform.

The more samples per second, the closer together the samples will be, giving a more faithful digital

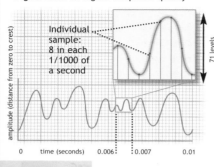

Digital audio at digital telephone quality

Individual sample: 8 in each 1/1000 of a second

71 levels

amplitude (distance from zero to crest)

0   time (seconds)   0.006   0.007   0.01

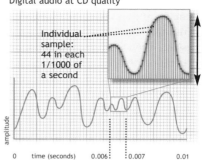

Digital audio at CD quality

Individual sample: 44 in each 1/1000 of a second

amplitude

0   time (seconds)   0.006   0.007   0.01

more is better
The more samples taken during each second, the more faithfully the original will be reproduced.

representation of the waveform and therefore a better quality sound. The length of the binary number that represents each sample is also important: the more bits per sample, the better the quality. CD-quality sound uses 44,100 samples per second and 16 bits per sample. But it is increasingly common for sounds to be sampled 96,000 times per second on a 24- or even a 32-bit scale, giving a sound quality superior even to that of CD.

Digital sound playback involves recovering the numbers and reconstructing the audio signal using a digital-to-analog converter (DAC). The reconstructed audio signal is amplified and sent to loudspeakers.

changing discs
The maximum duration of a standard CD and an MD is the same: 74 minutes.

CDs, minidiscs, and MP3s

The first digital sound format available to consumers was the compact disc. Philips first demonstrated their prototype in March 1979, which persuaded Sony to join in the development of a commercially available player. Eventually, in late 1982, the compact disc player appeared in the marketplace in Europe, and in the US early the following year. Other digital sound formats followed, including the popular MiniDisc (MD) from Sony in 1991. Both a CD and an MD can each hold up to 74 minutes of digital sound, despite the MD's smaller physical size and lower capacity for digital information. This is possible because an MD recorder processes the digital information using a code called Adaptive Transform Acoustic Coding (ATRAC). Each second of recorded sound is represented by fewer bytes of information on an MD than on a CD. This technique is called compression. The sound of an MD is almost as good as that of a CD, but compression does result in a small loss of quality. However, the MD format is a popular and convenient way to produce digital recordings.

music and the internet

There are other, similar types of compression, each one a complex mathematical procedure carried out by microchips in digital devices. These coding schemes, or codecs (coder/decoders), are all designed to reduce the number of bytes required to represent sound, while retaining as high a quality as possible. One of the driving forces behind the new codecs is the internet (see page 46), which has opened up new possibilities of sharing and buying music.

Transmitting music in digital form across internet connections is much quicker when the music is compressed, because it involves the transfer of smaller amounts of digital information. The most popular compression format, MP3 (MPEG Layer 3), was developed by the Moving Pictures Expert Group (MPEG), as part of an initiative to compress digital video. Tracks of digital music encoded as MP3 audio files can be played on computers or on MP3 players.

music by MP3
MP3 players use solid-state memory instead of a tape or a CD to store the music. They include software that transfers MP3 files from a computer into the player, and some also include utilities for copying music from CDs or web sites. It is also possible to create customized playlists.

digital images

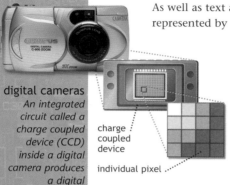

As well as text and sound, images can also be represented by collections of binary digits.

Digital cameras and scanners can produce digital images of real objects, while many other digital devices can produce or display sophisticated artworks.

To understand how digital photographs and scans are produced, imagine placing a grid of squares over a black-and-white photograph. Now imagine measuring how light or dark each square is. Give each measurement a number, list the numbers for all the

digital cameras
An integrated circuit called a charge coupled device (CCD) inside a digital camera produces a digital representation of an image.

charge coupled device

individual pixel

resolution

The size of an image, in pixels, is called its resolution. The higher the resolution, the more detailed the image will appear. At high resolutions, the pixels that make up a digital image (right) are not discernible. At lower resolutions, the pixels become visible. The image becomes "pixellated." A computer monitor typically has a resolution of 640 x 480 or of 1,024 x 768 pixels. So, an image ten pixels along each side appears as a tiny square in the corner of a monitor screen. A full-

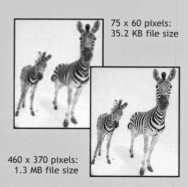

75 x 60 pixels: 35.2 KB file size

460 x 370 pixels: 1.3 MB file size

screen grayscale image at a resolution of 640 x 480 requires 307,200 bytes (one for each of the 307,200 pixels), or 300 kilobytes.

squares, row by row from top left to bottom right, and you have a numerical representation of the photograph. This list of numbers – in the form of binary digits, of course – is often called a bitmap.

The squares that make up a bitmapped image are called pixels – short for picture elements. Typically, the brightness of each pixel is measured on a scale from 00000000 for black to 11111111 for white (256 different levels). So, each level on this "grayscale" requires eight bits – one byte. A grayscale image ten pixels across by ten pixels down can be represented by 100 bytes – one byte for each of the 100 squares.

color images

Nearly all computers, digital cameras, and printers are capable of working with full-color images. The human eye senses color because it has three types of color-sensitive cell: one type reacts to red, one to green, and one to blue light. The eye can be fooled into seeing any color by mixing together only red, green, and blue light. Color printing, television screens, and digital devices take advantage of this: if you have ever looked closely at a television screen you will have seen that the pictures are composed of tiny red, green, and blue dots. From the normal viewing distance, the dots appear to merge, combining together to produce all the different colors that are seen in the full-color television picture.

Inside a computer, color images are composed of pixels in the same way as grayscale images. But each pixel now needs a number for the level of red, green, and blue present – three times the amount of information as for a grayscale image. The levels of

········· 00000000

shades of gray
The grayscale (left) runs from black (all 0s) at the top, to white (all 1s) at the bottom.

········· 01000000 (64)

········· 01100100 (100)

color bitmap
This bitmap image is displaying 24-bit color depth; the inset shows the individual pixels that make up the image.

········ 11011010 (218)

········ 11111111 (255)

brightness for each color are often represented by one byte of information. In this case, three bytes (24 bits) are required for each pixel – this is often referred to as a 24-bit color depth. Most computer monitors are able to display 24-bit color depth, also called true color, because more than 16.7 million different colours are possible (from 00000000 00000000 00000000 to 11111111 11111111 11111111) – more than the human eye can distinguish.

compressing images

Just like digital sound, digital images can be compressed. Again, the idea is to reduce the sizes of files without losing too much quality. Two common formats of compressed digital images are the Graphical Interchange Format (GIF, pronounced "giff") and JPEG (pronounced "J peg"). These formats are often used to display images on the internet, since in their compressed form they appear relatively quickly. The GIF format reduces the number of colors to a maximum of 256, requiring only eight-bit color depth. The JPEG format is a more complicated approach that takes into account peculiarities of human vision when compressing image files, and generally results in better quality images. It was devised by the Joint Photographic Expert Group (JPEG) and, like the popular MP3 format, which is used for digital sound, it actually refers to an approach to compression rather than a particular compression code.

saving space
Compressed bitmapped images take up less space inside a digital device. It also takes less time to transfer a compressed image from one device to another.

vector-based graphics

Bitmapped images are the only way to produce realistic digital representations of scenes or objects. Their drawback is that when they are enlarged, their pixels become visible as distinct squares. A useful way to represent images other than real scenes is to use what are known as vector-based graphics.

To see how vector-based graphics work, imagine you needed to produce a digital image that was a simple diagram of a circle. One way would be to paint the diagram on paper and use a scanner or digital camera to produce a bitmap. Now imagine instead that the image file simply contained the instructions for drawing the circle: the color of the circle, the position of its center, and its radius. This vector-based graphic uses a small number of binary digits, ensuring a very small file size.

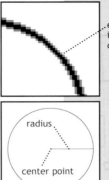

..each pixel individually defined

radius

center point

displaying images

A digital image can be held in any digital storage medium – for example, as a file in a computer. Normally, the file contains other information – in the form of Extended ASCII characters – such as the size of the image, the application (computer program) or device that created it, and the dates it was created and last modified. A digitized image can easily be manipulated. For example, special effects can be applied, such as blurring, coloring, embossing, or distorting. There are many image-manipulation programs available.

A monitor used to display digital images works in broadly the same way as a television screen: a beam of electrons called a cathode ray, produced at the rear of the tube, strikes pigments called phosphors on the back of the glass screen and causes them to glow. In the case of a monitor, a device called a display adapter converts the digital data into analog signals to generate the cathode rays.

vector factor
In the upper bitmapped image, the circle is defined by each of the pixels in its outline. In a vector image (lower) the inform-ation for the circle (radius and position of center point) is contained in a much smaller file and produces a non-bitmapped effect.

special effects
A wide range of special effects can be applied to digital images quickly and easily.

digital video

Once you understand how sounds and pictures can be digitized, understanding the principles behind digital video is only a short step away. Just as cinema films or television programs consist of a series of still images (frames) shown rapidly one after the other and synchronized with a sound track, digital video can be achieved by displaying a series of still digital images accompanied by digital sound. In its simplest form, this just consists of a stream of binary digits that represent the still images and the sound. About 30 MB (more than 200 million bits) are needed for each second of broadcast-quality digital video in this basic, "raw" form.

frame work
Conventional 16mm film (as shown here) is usually shot and played back at 16 frames per second. So an eight-frame strip contains the images to produce half a second of moving pictures. This illustrates the task facing any device required to produce high-quality digital video.

processing digital video

In practice, the bits comprising digital video are processed so that fewer bits are needed without a reduction in quality. There are many ways of achieving this, including recording fewer frames per second or using mono sound. Another method used in processing digital video for DVDs is to analyze groups of frames and only store significant frames along with the binary digits for parts of the picture that change from frame to frame. This procedure, devised by the Moving Picture Expert Group, is called MPEG compression.

three-dimensional digital worlds

Just as binary numbers can define two-dimensional objects in vector-based graphics (see pages 22–3), they can also define three-dimensional objects. In fact, entire "virtual worlds" can be defined and explored by using computer programs that can display a collection of virtual objects from any angle. Objects in a virtual world can be moved or rotated, and, with the aid of a virtual-reality headset, can be experienced in three dimensions. Game consoles make use of virtual worlds: some modern computer games present stunningly real environments for the players to explore.

early virtual-reality machine

storage

Whether binary digits are representing text, sound, images, or video, there is a variety of ways in which they can be encoded and stored. Computers store the bulk of their information in hard disk drives, which typically are now able to contain many gigabytes of data. Most computers can also read CDs or DVDs, which can store more than half a gigabyte. Some also have CD writers or CD rewriters, with which users can produce their own CDs containing multimedia. Many digital devices other than personal computers make use of the existing storage media used by computers, while others use storage media designed specifically for them.

hard disk
A computer's hard disk consists of aluminum or glass platters coated with magnetic particles. These particles can be oriented to show 0 and 1, allowing digital information to be stored.

platter

the read/write head writes data on the platter and also reads it

using digital technology

Representing text, sound, and images digitally opens up new possibilities for both organizations and individuals. The cost of even very sophisticated technologies has decreased dramatically over the past ten years due to new manufacturing techniques, which has brought digital technology to ever-increasing numbers of people. In technologically developed countries, most people now have some form of access to a personal computer – the all-in-one, universal, digital device. Being digital, computers can easily connect together to form networks through which hundreds of millions of people worldwide connect their computers to the ultimate network: the internet. People use the internet to communicate via email cheaply and easily, to advertise their products, to stay informed, to carry out research at all levels, and even to shop and be entertained. Other digital technologies with widespread appeal are mobile communications and digital broadcasting, which are both gradually superseding their analog counterparts.

digital desk
Mixing boards in today's recording studios take full advantage of digital technology. They can mix up to 96 audio channels, eight surround-sound channels, and produce high resolution graphics. They can also combine digital and analog inputs and outputs.

the personal computer

A laptop or desktop personal computer is an all-around digital device with which you can create and print text documents; create, manipulate, and print images; organize information in databases; communicate across networks; play games; watch and edit digital video; record, edit, and play back digital sound, and much more.

Inside a computer's case, most of the components are attached to a large circuit board called the motherboard. Perhaps the most important of these components is the central processing unit (CPU), a powerful microprocessor

processing power
The modern computer is a processing tool, which first allows the material that is to be processed to be input. The internal components of the device perform the processing, and once the results are complete, allow them to be output.

input	storage and processing	output
camera	hard drive	monitor
mike	CPU	printer
keyboard		
mouse		
scanner	RAM	speakers

that carries out most of the processing of digital information. The CPU follows binary-coded sets of instructions (programs, also called applications) that are temporarily stored in random access memory (RAM). Programs and files are loaded into RAM from the computer's main storage medium, the hard disk, as they are required, and output is sent to a variety of devices, including the monitor, printer, and loudspeakers. Output depends on what is input into the computer, via any of a wide range of input devices including the keyboard, mouse, scanner, digital camera, and microphone.

early ancestor
The Difference Engine, a calculator designed by Charles Babbage (1791–1871), is an early precursor of the modern computer.

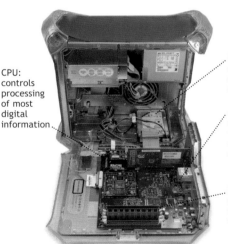

CPU: controls processing of most digital information

hard disk: stores computer's applications and files

motherboard: circuit board to which most components are attached

RAM: in which applications and files are stored when in use

main parts
There are many different types of personal computer, but they have much in common. For example, every personal computer has a CPU, RAM, a motherboard, and at least one hard disk.

operating systems

The most important program run by the CPU is the operating system. This is loaded into RAM whenever the computer starts up and then controls the flow of all digital data. The two most popular operating systems are

icons

Both the Macintosh and the Windows operating systems use icons to allow the user to access applications, as well as the folders and files, easily.

application

folder

file

Windows, produced by Microsoft, and the Macintosh Operating System, by Apple. Both provide a graphical user interface (GUI, pronounced "gooey"), which represents files and programs as small pictures, called icons, on the monitor. This method makes it easy to access and organize the contents of the hard disk.

expanding and connecting

connecting

Digital information passes between the CPU and external, or "peripheral," devices via cables that connect to the motherboard through sockets called ports.

The CPU is connected to various devices inside the computer including separate microprocessors that carry out specific tasks. Circuit boards called expansion cards are usually plugged into the motherboard, and connected to the CPU via metal tracks called buses. One expansion card digitizes analog sound signals, while another produces output for the monitor. This leaves the CPU with more processing time to carry out the tasks demanded by the programs that are running.

External devices can be connected to the CPU using a variety of standard cables, which plug into sockets called ports.

While the personal computer is a digital device, it sometimes needs to produce analog output signals. For example, to hear sounds produced by the computer, an amplifier and loudspeakers must be connected. These are analog devices, so the sound card in the computer converts the digital sound signals produced by the CPU into analog signals for the amplifier. However, most external and internal devices to which the CPU connects are digital. Links between digital devices – digital networks – underpin modern business, and have integrated seamlessly into many other areas of our lives.

Small Computer Systems Interface (SCSI), "scuzzy," is faster than a parallel cable.

A parallel cable can transmit one or more bytes at a time.

Universal Serial Bus (USB) has greater capacity than SCSI and parallel cables.

digital networks

Connectivity is one of the main strengths of digital devices. A network is formed whenever two or more digital devices are connected together. Digital networks are increasingly common in the modern world: cash machines are all connected to a bank's central computer; a typical office has several personal computers and a printer connected together; most supermarket checkouts are networked to central computers that keep track of stock levels; and most telephone calls are made across digital networks. Digital devices can form temporary networks, as well as those that are more permanently connected. For example, entries in an electronic diary or personal digital assistant (PDA) can be transferred to a personal computer through a docking cradle or even via an infrared link. In these cases, the network only exists for as long as it takes to transfer the information.

net cash
Bank customers make use of digital networks when they use cash machines.

air net
Networks can be temporary or permanent, and can be made through metal wires or fiber optic cables – or through the air, using infra-red or radio waves.

handshaking

Information that passes between devices over a digital network must follow rules called protocols. The use of these protocols ensures that the information is delivered reliably across a network. When devices first connect to a network, they communicate with each other to ensure that they are using the correct protocol. This process is known as handshaking. For example, the sounds made by fax machines when they first connect include high-pitched beeps, which

carry coded information about the two machines. Once the handshaking is complete, the network between the two machines is established, and information can pass effectively between them.

agreeing protocols
Just like people, digital devices need to establish mutual understanding before they are able to communicate efficiently.

local area networks

The most common type of network is the local area network (LAN) – the sort of network that connects personal computers and printers together in an office, for example. In large organizations, several offices might be on the same LAN, sharing files between departments and being able to communicate via electronic mail (email).

The most common protocol used on a LAN is called Ethernet. Each device connected to an Ethernet network has a unique address that identifies it. Ethernet cables allow high speed transmission of digital information. Computer files are sent across the network contained in small groups of bytes called frames. Each frame contains a file or part of a file being transferred between two devices on the network. It also contains the address of the device that is the destination of the file, and other information necessary for its successful delivery. The standard connection used in Ethernet LANs – particularly those in offices – is an RJ-45 cable, which looks like a fat telephone cable.

cable speed
These RJ-45 Ethernet cables (above) are capable of transmitting one megabyte of data each second.

servers

In most LANs, at least one of the computers is a server. The purpose of a server is to store files that can be made available to people who are connected to the network. Some of the files will be applications, which all the connected users can run without having to have a copy stored on their own PC. This also means that everyone in the same network will be using the same set of applications. Often, servers are simply desktop computers that run server applications. LANs that have no server are called peer-to-peer networks – the files and applications can be on any of the computers and are shared between all the machines connected to the network.

Not all files held on a server are available to all the network's users: different users have different access privileges. For example, only the accounts department might be able to access information about payroll.

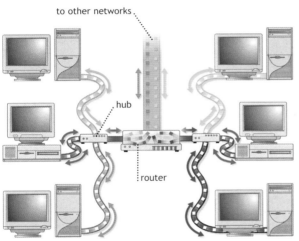

to other networks

hub

router

ethernet network
Many Ethernet networks include devices called hubs, which allow easy connection of new computers into the network. Large LANs can be subdivided by using routers, which control the flow of network traffic. A router can also help to connect one network to others, forming an internetwork.

Sometimes, a user can gain specific privileges by entering a password. Collecting email is an example. In a typical LAN, email messages have their own server, called a mail server. This holds email messages from one network user to another, and which can be accessed only by the person for whom they are intended. Once the correct password is given to gain access to a particular location (called a mailbox) in the mail server, special e°mail software is used to communicate with the mail server, and to collect (download) and display the relevant messages.

connecting networks

networked computer

routers

internet server

A local area network with many users can become congested, because only one frame of information can be released onto the network's cables at a time. Large LANs therefore have several servers plus a general mail server. Additional devices called routers direct network "traffic" to the servers. Nearly all LANs are themselves connected to various other networks, so forming internetworks. An individual LAN is often referred to as an intranet.

Routers run programs that plan a route for each frame of information via several other routers. This is particularly useful if some parts of a network are congested or out of action, because the information can be routed around the problem. Different networks may work according to different protocols, and a device called a bridge is used at any connection between networks, to translate between the different protocols.

internetworking
Routers allow two or more networks to connect together, directing network traffic and translating the different protocols.

TCP/IP

To ensure that networks can communicate, protocols are used that can work on any network. One example is the Transfer Control Protocol/Internet Protocol (TCP/IP), which is a family of protocols enabling file transfer across many different internetworks. TCP/IP forms the basis of the internet.

Any internetwork has a system of high-speed digital links called a backbone. A device called a gateway allows traffic to pass between LANs and the internetwork's backbone, and forms a connection known as a network access point. The internet, then, owes its existence to servers, bridges, routers, and gateways. Together, these can join networks that are on opposite sides of the world.

key points

• Information passing between digital devices follows rules called protocols.
• Computers and printers on a local area network most commonly use the Ethernet protocol.

the internet

Just about everyone in the developed world is aware of the existence of the internet – the ultimate internetwork, which connects computers in all continents of the world. The wealth of services offered by the internet is impressive – home shopping and electronic banking are obvious examples. The internet creates unprecedented opportunities for international communication and collaboration, for education, and for the dissemination of information. Noone has ultimate control of this worldwide network – a fact that brings with it freedom of expression. This can have both positive and negative consequences. For example, it means that the internet can

getting connected

Large organizations have their own dedicated connection to the internet's backbone. Most individuals and companies, however, connect to the internet via servers owned by companies called internet service providers (ISPs). An ISP provides a network access point (NAP) via a special server called an access server, which acts as an internet "point of presence" (POP). Individuals or small organizations have permanent, always-on connections, or connections via dial-up modems. Whenever users are connected to the ISP's server, they are also connected to the internet. In this way, an ISP provides a way of accessing information from any other network on the internet.

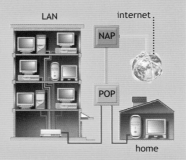

be used to expose immoral practices or it can enable people to find new ways to exploit others.

modems

A modem (modulator/demodulator) converts digital signals output from a computer into a form that can be transmitted across a telephone line. When a modem dials a telephone number, devices called switches connect that modem to another modem at the ISP. A modem produces a high-pitched sound called a carrier signal. It continuously changes (modulates) certain characteristics of this signal to represent groups of binary digits. The modulated tone is transmitted across the telephone line in the same way as the sound of your voice. Another modem, connected to the access server, demodulates the signal, recovering the binary digits. Information passes in both directions in this way, enabling dial-up users to send the server requests for information from the internet and receive information from servers run by other ISPs.

modems
A modem encodes digital information – text, images, data, video – from a personal computer into an analog sound signal. The signal passes across a telephone network, and is then converted back into digital information by another modem. This process occurs irrespective of the direction of travel of the data.

Most modems can transfer up to 56 kilobits per second (Kbps). This is close to the limit of the speed of information transfer via a conventional telephone system. One kilobit is equal to 1,024 bits (just as one kilobyte is 1,024 bytes), so in principle 57,344 bits (7 KB) could pass between two modems every second. Modems rated at 56 Kbps

When, in the 1940s, **Vannevar Bush** (1890–1974) developed the idea of the "memex" – a storage and information-retrieval system – part computer, part database with a screen and a keyboard, he predicted the invention of the personal computer. He also described an early version of the internet with hypertext linking.

information motorways

The speed of information transfer across a network depends on bandwidth – the volume of data that can be sent in bits per second. The conventional telephone service that links an ISP with its dial-up customers is described as a narrowband connection. All internet connections between networks are broadband connections: they can achieve transfer rates of up to several hundred megabits per second (Mbps). An emerging technology, called dense wavelength division multiplexing, makes it possible to transfer 400 gigabits (409,600 Mbps) per second through a single optical fiber. You can think of internet backbones as information highways, carrying large volumes of fast-moving traffic, while an individual's connection to an ISP is the equivalent of a local road. As the availability, popularity, and content of the internet grows, more people want to have the highway at their door. So, ISPs have gradually begun to offer broadband access to the internet.

make use of a technology called asynchronous transfer mode (ATM), which means that information is transferred more quickly downstream (from the ISP's server to the customer) than upstream (the other way). The actual rate of transfer depends on several factors, such as how many users are connected to the access server.

broadband options

Several options are already available to ISP customers who want to be able to collect information more quickly from the internet than with a dial-up modem. These technologies that transfer information at between 64Kbps and 1.5Mbps are sometimes described as wideband, but most people use the term broadband. All these options provide more bandwidth than dial-up modem connections, as they do not make use of conventional telephone technology. Early telephone systems carried sound as electrical signals, which were analog audio signals, which are a direct copy of the sound waves spoken into the telephones. However, most modern telephone networks transmit digital signals. These are carried on a stream of 64 kilobits (8KB) per second. During a telephone call, the sound of your voice is digitized either at a box in the street or at the telephone exchange. The digitized sound is transmitted – along with millions of other signals – across high-speed digital networks. It is decoded back into an analog signal at the receiving exchange, or at a roadside box, before reaching the receiving handset.

old and new
ISDN and DSL achieve broadband access to the internet via existing twisted pairs of wires – the basis of the original analog telephone networks of more than a hundred years ago.

ISDN

One of the oldest of the emerging broadband options is Integrated Services Digital Network (ISDN), which makes use of the existing telephone line, but in an all-digital way. Customers whose telephone signals are digitized at the exchange, and not at a roadside box, can send digital signals from their computer all the way to the exchange. At the exchange, the digital signals bypass the analog-to-digital converter, and continue as digital signals to their ISP. In this way, each telephone line can transfer digital signals at 64 kilobits per second – faster than a dial-up modem. All households have at least two lines – one is normally redundant – giving a possible 128 kilobits per second. By installing more telephone lines, several ISDN channels can be combined, giving even greater bandwidth.

video conference
One of the first uses of ISDN was to connect people in conferences across continents.

DSL and cable modems

Other popular examples of broadband technology are DSL (Digital Subscriber Line) and cable modems. Like ISDN, DSL makes use of the wires present in telephone cables. Telephone calls use only a tiny part of the potential bandwidth of these wires. DSL makes more use of this potential, and transfers digital information coded into high-frequency electrical signals. It uses the same approach to modulating its signals as dial-up modems, but the bandwidth is much greater. Cable modems work in a similar way, but the information is transferred along television cables, or as digitally encoded laser light through the fiber optic cables supplied by cable television companies.

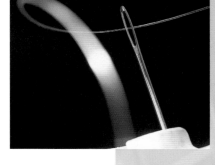

Both cable modems and DSL are capable of transferring several megabytes per second, but domestic versions typically allow users to download at 512 kilobits (64 KB) per second. Both have other advantages over ordinary dial-up modems. Dial-up users have to wait every time while their modems handshake with the ISP's modems. They also incur telephone charges whenever they are

signal conductor
Fiber optic cables are composed of hundreds of thousands of hair-fine individual fibers. This makes the cable flexible and able to carry vast quantities of data along an extremely fast connection.

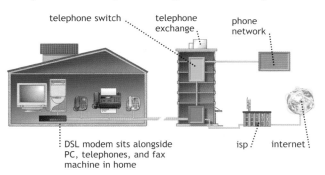

telephone switch
telephone exchange
phone network

DSL modem sits alongside PC, telephones, and fax machine in home

isp   internet

fast connections
High speed DSL provides fast internet access, yet uses the same cables as ordinary telephones and fax machines.

digital address
Digital information is sent across the internet in packets, each containing codes that direct it to the correct address.

packet codes
Shown here is the digital code for a packet of information. Colors represent: its total length (magenta), the position of its information in the complete document (yellow), its protocol (orange), the source IP address (purple), and the destination IP address (blue). The information in the packet is shown in black.

online – connected to the internet. Cable and DSL users retain their connection to the ISP even when their computer is switched off, normally for a flat monthly fee.

packets

In whatever way someone connects to their ISP's servers, each file they exchange is divided up and coded into small groups of binary digits. Each chunk of digital information is called a packet and contains the address of the server for which the information is destined and other binary digits help the internet's routers ensure the information reaches its destination. The address of a server is called the IP address, and consists of 32 binary digits. For human convenience, it is normally written as four numbers separated by decimal points. In their true binary form, IP addresses form an essential part of every packet sent across the internet. They are assigned by an organization called the Internet Network Information Centre.

Transmitting computer files across the internet in packets means that the various fragments that make up a single document may travel quite different routes to their destination. This is a very effective way of using the available bandwidth of different sections of the internet. Every packet has a number, which identifies the place of

```
1101101011110100111011010110110110111011100110111011110110011101
0101110110011101010111101101011010011101101101101011101101101101
1101111001011101101101101101101110101001011100000110101011010
100010100000110101010111110011010111100000100101011111111111
110101010100101010101010101011111111111111111111111100000000
000101000000000000000000111111100101000000000000000000000000000
00000011111111111000000000011111111111111111111111111111110000000
001010101011100110010101010101010101010101010111111111010
0000001010101110100001010110101010101111111010100110101011001010
00000000000010101010101111111111101010000100100111010010000
000010101011010111000010111110101000001010110101000100101111111
11101010100010100001010110101011101010100101000000000000001010
1010101010101010011011010100100000010111111111010101010101010111
11111010101000000010101010001111111110101001111111110000000000101
0101010101011010100000011010011010101011010101010101010001010110101010
```

the information it contains in the file as a whole. Using packet-switching and the system of IP addresses, users can exchange any type of document or program with any server that is connected to the internet.

email and web pages

One type of server run by every ISP is the mail server, which stores email files. Normally, an ISP has one mail server that receives emails for its customers, and one that sends customers' emails. Each mail server has its own IP address. Email allows people all around the world to

incoming mail server of recipient ·.   ISP's mail server .

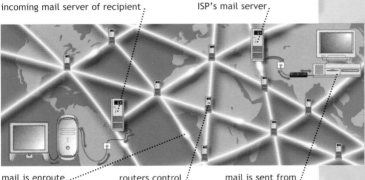

mail is enroute ·········   routers control :   mail is sent from :
across the network   the flow of traffic   this computer

communicate with each other quickly and conveniently. Hundreds of millions of emails are sent across the internet every day for countless different reasons: friends and relatives send them to keep in touch; businesses send them to contact actual and potential customers; people exchange computer files coded into email transmissions as "attachments." Email messages are also the basis of "newsgroups." These are basically internet bulletin boards published on "news servers."

Another type of server, called a web server, holds documents that can be read by anyone connected to the

sending messages
Emails pass first to the mail server at the sender's ISP. They then travel, in packets, via the internet and directed by routers to the incoming mail server at the ISP of the person who is to receive the message.

internet. Applications called browsers are used to display these documents in the form of "web pages." The most popular of these browsers are Netscape, written by the Netscape Communications Corporation, and Microsoft's Internet Explorer. Fortunately, users do not have to input the IP address of the relevant server to request the required web page document. Instead, a keyword called a domain name is assigned to each server connected to the internet. So, for example, 158.43.129.68 becomes www.dk.com. The www part of the domain name indicates that the host is a web server.

HTML
Web pages are written using a language called HTML (inset). Codes called tags are used to format text and images, so that the document will produce a web page when viewed with a browser. Tags also denote the embedded hypertext links that are perhaps the most important feature of the web.

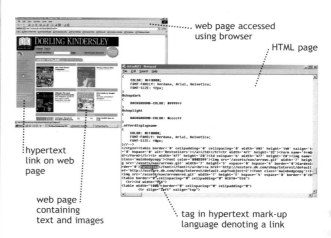

web page accessed using browser

HTML page

hypertext link on web page

web page containing text and images

tag in hypertext mark-up language denoting a link

hyperlinks

Embedded in a web page are references to other files, which may be located anywhere on the internet. These references are called "hypertext links," "hyperlinks," or simply "links." Accordingly, the internet protocol used by browsers to access web pages is called HTTP (the Hypertext Transfer Protocol), and a document that defines how a web page will look is written in a computer language called HTML (Hypertext Markup Language).

java and applets

A very large number of web pages include elements of a computer programming language called Java. While HTML does little more than display text and images, Java is used to create "scripts" within HTML documents that enable web page designers to include an element such as background music, an animation, or live digital video, which is displayed within the web page, and multiplayer games. Scripts also allow a Java-enabled browser to open types of documents that the browser alone could not open, such as particular types of spreadsheets. Java programs embedded in web pages are called "applets" – a shortened version of the word "applications."

world wide web

Anyone with experience of the internet has viewed web pages and probably explored the links that they contain. To follow a link, you use a mouse to move a pointer over certain parts of a web page displayed on a monitor screen, and then click the mouse button. This causes the browser on your computer to send a request to the web server holding the linked file – typically another web page.

"The world has arrived at an age of cheap complex devices of great reliability, and something is bound to come of it."

Vannevar Bush, 1945

The links within an HTML document are in the form of a uniform resource locator (URL), which contains the destination server's domain name, as well as the protocol required to open the requested document. A server responds to a request by sending a copy of the linked file to the computer you are using. This linking of web pages on the internet effectively forms a complex and ever-changing virtual mesh of information. This is called the World Wide Web (WWW), usually known as the Web.

ISP customers often produce web pages themselves, and transfer them to the web server – a process called uploading. The presence of a document on a web server makes it available to anyone else connected to the internet, wherever they are, and is therefore an effective way of publishing information.

web sites

A collection of interrelated web pages, contained in the same folder in a web server, is called a website. All websites have one specific main page, called the homepage, which provides a way into the site and provides links to the site's other pages.

information breakdown
Most websites are a hierarchy of information. The homepage contains links to areas of interest on the site, and the other pages link back to the homepage.

homepage

link to page

page in site's hierarchy

There are many kinds of individuals and organizations who choose to publish their information on websites, for many different reasons. For example, universities explain their research and try to attract new students; businesses advertise their services; government and nongovernment organizations disseminate public information; transport companies make their timetables available; news media vie with one another to be first with the news, and individuals publish photographs of themselves and their families. Other services can be accessed via web pages, including online banking.

snap-shot
Individuals' web pages often contain images of their families. This can be an effective way of keeping in touch with distant relatives.

searching

Large websites have their own search facility, through which internet users can locate files of interest held on the site's servers. The user sends a keyword, via their browser, to an application called a search engine. This examines the contents of each HTML file held on the server and compiles a list of any that include the

British computer scientist **Tim Berners Lee** (b.1955) helped to weave the World Wide Web. In 1990, while working at a scientific research center in Geneva called CERN, he invented HTML and put forward his vision of how the web might work. Before the existence of HTML, the internet had only been used by academics and military agencies.

keyword. This takes a fraction of a second, and the results are published on the web server in HTML documents. The internet user's browser automatically opens the results page, and displays links to the pages listed there. Some search engines do not restrict searches to documents on their own server. Instead, they "crawl" the web – automatically following links from web page to web page and storing copies of the HTML documents of the pages they visit. By doing so, they build up a database that contains copies of many millions of web pages. When someone sends keywords to one of these search engines, it performs a search on all the documents in its database. In effect, this is a search of documents held on many thousands of different servers across the entire internet.

streaming

Using a browser, internet users can download files that may be useful or interesting to them. Many people like to download applications, such as computer games. Others are more interested in downloading text documents or digital sound, image, or video files. Most of these files cannot be processed until they have been downloaded in their entirety. But certain types of digital sound or video files can be processed as they are received. This is called streaming. Links embedded in the web pages lead to the server that holds the relevant files. Often there is a choice of links on the web page: streaming files of different quality are matched to the bandwidth of the user's connection. The higher a user's bandwidth, the better the quality they can receive.

MP3

Whether downloaded in their entirety or received in a streamed format, most computer files – and in particular, sound and video files – are compressed. This reduces the time it takes to download them.

fast and slow
Using a dial-up modem, large files can take a long time to download, particularly if they have not been compressed. This is one reasons why broadband connections to the internet are growing in popularity.

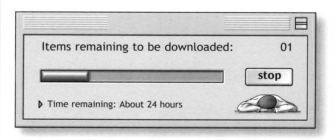

Items remaining to be downloaded: 01

[   ]   stop

▶ Time remaining: About 24 hours

The most popular compression format for sound files downloaded from the internet is MP3 (see page 19), which can deliver sound with near-CD quality using as few as 64 kilobits for each second. At the same transfer rate, uncompressed digital sound on an ordinary telephone line, for example, has a far inferior quality. MP3 has become very popular. Internet users can download an MP3 file from any of thousands of web pages, and listen to them using their computer. Many internet users quickly build up large collections of MP3 files on their hard disks, and have portable MP3 players into which selections of files can be transferred. Some musicians convert their own music into MP3 files, and make it available on their web sites.

player controlled by buttons on watch face

napster

The MP3 music format has changed the way a large number of people listen to, buy, and share music. One application that has played its part is called Napster, which allows people to share music files online. Millions of people have signed up to the service run by the Napster company, which holds a database of its members together with details of which music tracks they have available on their own computers. Music can then be transferred digitally between any members. This service was originally offered free – a move that was popular with neither the existing music industry nor copyright lawyers. After long and much-publicized court proceedings, it was agreed that Napster would charge for its service in the form of a subscription, and that a percentage of the profits would be paid to the owners of the music's copyright. Napster challenged traditional copyright laws, and forced record companies to look at alternative methods of protecting their products, which has enabled some companies to begin to sell their own music online. To do this, and to sell other goods online, people must be able to carry out financial transactions via the internet.

MP3 watch
MP3 files can be downloaded from the internet and then played back on computers or digital devices such as a watch containing an MP3 player.

note the copyright
Napster has made the music industry realize that to control copyright violations, music must be protected before, during, and after its release.

viruses and worms

A particular threat to internet security comes from certain types of information transmitted on it. Any type of file can be sent to any computer connected to the internet, typically via email. Some files are small programs that, once present in the computer of an internet user, can cause some serious problems. Such programs insert their own binary code into the programs already in the computer. This can lead to the destruction of information, or to other problems that may stop the computer working. These programs are called viruses, because they infect a computer and normally work by rapidly replicating themselves. Viruses can spread across networks, and can cause a great deal of damage and annoyance as they do so. Similar to viruses are worms, which are more self-sufficient; they work without having to hijack other programs. Worms can quickly spread across an entire computer network, typically by programming email applications to forward copies of the worm to everyone listed in a computer's email address book. Warnings are published on the internet as soon as a virus or worm has been detected. Many anti-virus programs, which are designed to prevent known viruses from infecting computers, are commercially available.

virus invasion ·· wall of host cell
Both computer and biological viruses, (shown above), penetrate defenses.

e-commerce

The increased use of the internet has led to a variety of new opportunities for business. One of the most important innovations is the capacity to transfer money on the internet by using a system of electronic money transfer, which already existed before the rise of the internet: the credit or debit card. The capacity to transfer money over the internet allows people to do their banking online, to do their shopping at "virtual shops"

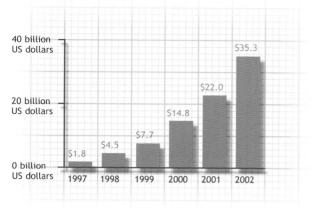

e-commerce
Over the past few years, the amount of money spent buying goods and services online has shown a dramatic increase, and is predicted to continue to rise.

on retailers' websites, or to pay for software that they download to their computers.

Many businesses now also regard the internet as an attractive location for placing advertisements. Websites that present exciting content generally attract more internet users, and can therefore charge higher rates for companies to advertise on their web pages.

internet security

Many commercial websites make use of small computer files called cookies, which are downloaded to the hard disk of an internet user's computer and contain personal details about that person as well as identifying their computer. A cookie will usually contain the name and e-mail address of the person, along with details of when they visited the website. Other information about the website's visitors – such as a list of all purchases they have made and their method of payment – may be stored on the company's servers. This can bring convenience to internet users

security symbol
Whether you are using Internet Explorer or Netscape Navigator, you will see a padlock symbol at the foot of the window when you have entered a secure site.

key points

• The internet, is a worldwide network of computers.
• Users of the internet can exchange email messages, and can access the World Wide Web, a huge collection of interlinked web pages.
• Many services are available via the World Wide Web, including e-commerce and streamed digital audio and video.

because they will not have to enter all their personal details each time they want to shop online. But it also raises important questions about privacy and security.

Just as people can tap into telephone conversations without the caller knowing, so information transmitted across the internet can be intercepted. Some internet sites therefore publish certain pages on a "secure" server. Communication between a secure server and an internet user's computer is coded according to a number called a key, which is used to encrypt the details that are sent over the internet, so that anyone intercepting the information would be unable to make sense of it. This is called cryptography. It provides a secure way to make purchases online by using, for example, a credit card.

crackers

Another threat to the security of information on the internet comes from crackers: people who use computers to access secure information. Crackers bypass password controls, or find other ways to exchange information with secure servers and use that information to cause damage. This process is called cracking, and can compromise the security of information held on commercial or government networks. Many people mistakenly use the word hacker to describe a cracker. Hackers do not compromise security: they are expert programmers who share the results of their work via the internet. A large number of popular computer applications – most of which are free to use – have been designed by teams of hackers.

cybercrime
Despite safeguards such as passwords put in place by internet service providers, the illegal acquisition of protected information on the internet is still a problem.

entering a password provides only partial protection

digital airwaves

The mobile, or cell, phone provides the most public evidence of the spread of digital technology today. And, as is the case with most digital communication,

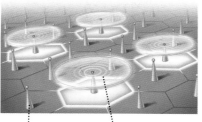

base station mast : area of transmission

it is based on networks. Cell phone networks are divided into cells, that are about 10 square miles (26 square kilometers) in area, and each contains a base station with a radio mast. When you switch your phone on, it logs on to the network's control office via the base station of the cell in which you happen to be. When a call is received for you, the switching office tells your phone which frequencies to use, and once your phone and the base station switch to those frequencies, the call is connected and you can start speaking.

As you move towards the edge of your cell and closer to another, the base station in the cell that you are approaching sees that your phone's signal strength is increasing. The two base stations coordinate with each other, and your phone gets a signal to change frequencies. At a particular point, your phone is switched to the new cell and you simply continue with the call.

New mobile telephone services – called Third Generation (3G) systems – provide a means to access the internet directly

digitizing the voice
The mobile phone converts the analog signals of the human voice into digital signals, which can be received and transmitted by the nearest base station.

DAB
Analog radio signals suffer interference caused by obstacles and weather conditions. However, Digital Audio Broadcasting (DAB) uses these effects as reflectors creating multipath reception conditions to make most use of receiver sensitivity. DAB always selects the strongest regional transmitter, so you'll always be at the focal point of incoming signals.

from 3G-enabled mobile phones. Digital transfer rates across 3G mobile networks are up to 384 Kbps when the user is stationary, and 128 Kbps when on the move. This is fast enough for telephones to include small screens on which live digital video of the caller can be displayed. Mobile video phones, along with a wide range of other hand-held, 3G-enabled devices, will become common.

radio and television

Nearly all radio stations produce their programs using digital technology. Many, however, still output these programs using analog signals. These signals are transmitted by continuously varying the strength (amplitude modulation or AM) or the frequency (frequency modulation or FM) of a signal.

Increasingly, however, radio stations are broadcasting their signals digitally – modulating high-frequency radio waves in the same way as a modem modulates an audio signal. The digital sound is sent in packets of around 200 Kbps of binary-coded sound. Using digital sound as the basis of radio broadcasting makes it ideal for transmission via the internet. Many radio stations now transmit programs by publishing streaming files on the internet, as well as broadcasting them as digitally encoded radio waves.

Many television companies – whether they provide programs through radio transmitters, communications satellites, or optical fibers – have also begun to broadcast digital signals. This has led to new technologies, including pay-per-view – where certain programs are only available to paying viewers – and to video-on-demand, which allows viewers to watch television content whenever they wish.

digital broadcasting

Digital broadcasting has advantages over analog technology, for example, by sending error codes with digital signals to ensure clear reception. Additional information – such as song titles and program schedules – are included in a digital radio signal, and can be displayed on digital radio receivers. Some digital television services also offer a host of extra services, including simultaneous viewing of more than one channel and the inclusion of interactive features. One of these would be video-on-demand, where movies could be chosen and viewed at will. Much development is still required, but one method relies on a large database of movies. For this to be used, a very fast network is needed to send the movie through the cable television wires. Also, as movies are transmitted compressed, a device is needed to decompress them, and show them on a TV screen. This could by done by a set-top box containing the memory to hold an entire movie, and to play it.

main channel

second channel can be viewed simultaneously

channel options
Some digital television services allow two windows to be displayed, which can be resized, and some also offer the facility to pause live television.

the digital future

Advances in digital technology have enhanced many areas of life for large numbers of people. But those same people want faster, cheaper, more powerful, and more integrated technologies that are also easier to use. It is these desires that ultimately fuel the digital revolution. The pace of that revolution is such that it is possible to make predictions of short-term developments fairly reliably. Prediction beyond the next few years, however, is uncertain. For example, there is no consensus about the future of the personal computer (PC). It is thought by some that the advent of supersmart alternative devices will make the PC redundant as a separate computer. These people foresee a longer-term future in which digital technology fits seamlessly together, helping us to communicate, learn, work, shop, and enjoy – a future in which we are controlled and amused by digital technology. Whether or not this vision of the future is realized depends on many things – not least the desires of the people who will be able to afford it.

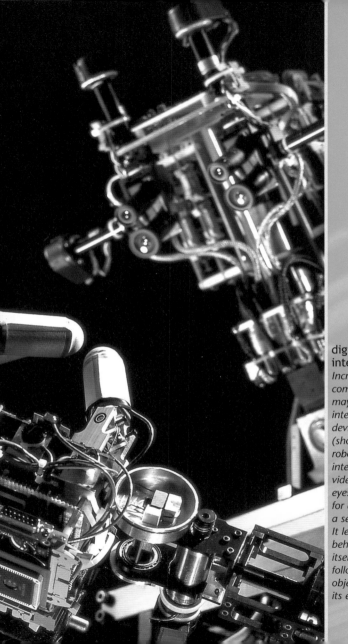

digital intelligence

Increased computing power may lead to more intelligent digital devices. COG (shown here) is a robot with limited intelligence. It has video cameras for eyes, microphones for ears, and even a sense of balance. It learns simple behaviors for itself, such as following moving objects with its eyes.

convergence

Digitizing text, pictures, sounds, and video is a way of expressing all of these separate elements in a single common language, made up exclusively of binary 0s and 1s. Many digital devices take full advantage of this universal form of expression. A modern digital personal organizer, for example, can also be a mobile phone, a digital camera, a digital sound recorder, and a fully functioning computer.

on the move
One of a rapidly increasing range of all-in-one mobile devices, this "communicator" is a mobile phone, an address book, and a hand-held computer.

merging technology

Because digital information exists solely as 0s and 1s, telephone services, digital broadcasting, and internet services can all be delivered through the same network at the same time. Technologies that were once separate have begun to merge, taking on each other's roles. For example, millions of people now make low-cost telephone calls using the internet. This IP (Internet Protocol) telephony is made possible by digital sound streamed across the internet. Many people also listen to music online, even purchasing and downloading their favorite tracks.

The DVD is another example of convergence within the digital world. DVD discs can be used to store large amounts of any kind of digital information – typically films, super-quality sound, or interactive encyclopedias – and can be read by a personal computer or a DVD player.

internet and television

The internet and television are the two technologies that are most likely to undergo complete convergence in the near future. This process has already begun with the

Home Theater Personal Computer (HTPC), which offers a wide array of television and internet functions.

Most cable television subscribers already have the option of broadband access to the internet via a cable modem. Some digital television companies offer internet access from a television set, and many television companies make their news programs available on their own websites, as streaming video-on-demand. In fact, most technology experts feel that, in a few years, television will become subsumed under the internet, because – in principle at least – the internet can do everything a television can do, and more besides.

new view
By offering home shopping, viewing on-demand, and access to the World Wide Web, digital television blends TV and the internet into one.

a future device

The HTPC will probably evolve into a device offering greater convergence of technologies and fulfilling the role of a television that provides more choice of what

htpc: sound and visionary

The Home Theater Personal Computer (HTPC) demonstrates the way the internet and television are converging. Anyone who owns an HTPC can watch high-definition digital television or DVDs (on large, crystal-clear, digital screens), browse the internet, listen to music, edit digital video, produce CDs, and send and receive email – all from their own living room. At present, however, the HTPC is very expensive, and therefore available to only a few people. It is also quite cumbersome, uses a lot of electrical power, and is complicated to set up. In a few years, the HTPC will probably evolve into something less power-hungry that takes up less room in a house, and configures itself automatically.

key points

• Developments in digital technology in the short term are relatively easy to predict.

• Technologies that were previously distinct, such as television and the internet, are now converging, and soon homes and businesses will use fully integrated digital technology.

to watch and when to watch it. Content delivered by future media companies will have to adapt – more will have to be available on-demand, for example, home entertainment and communication.

We will also use this device to access digital music stored on a computer elsewhere on the planet. We will pay our bills through it, order takeout food, and make appointments with the doctor and the hairdresser. It will allow us to access, and publish, information on the internet. We will communicate with the device by speech and access it from anywhere by using handheld devices that beam digital information to satellites. In fact, the device need not be in one single location in the home, but could be a network of smaller units in different rooms, each with a large, clear, flat screen, loudspeakers,

future home
Digital technology is likely to become increasingly integrated into our homes, accessible from every room easily, efficiently, and cheaply.

and a hidden microphone. To some this is a vision from science fiction, but there is no real technological barrier to it. Nevertheless, if this multipurpose device is to become an integral part of our lives, advances will have to be made in several digital technologies. Also, the device will have to be cheap enough for most people to afford – and enough people will have to want to share this vision.

the near future

Developments in digital technology have been extremely rapid since microprocessors started to appear in consumer products in the 1970s. The pace of change is likely to remain fast, urged on by consumers, researchers, investors, and "content providers." As we move towards an increasingly integrated digital future, many of the technologies that are at present found only in research laboratories will quickly be implemented in devices owned by millions of people.

magnetic memory

When personal computers became available in the late 1970s, they typically had a few kilobytes of RAM. In the year 2001, a typical desktop personal computer had 128 MB of RAM. The range of storage options has also increased rapidly and the price of these items has dropped massively in real terms, encouraging their use in many devices other than PCs. Even so, the relatively low capacity of modern digital storage options and the way they use power must be overcome before digital convergence can become a reality.

For example, the integrated circuits that make up RAM lose any information stored in them when the power is turned off. This means that information must be loaded back into RAM each time a device is turned on, and that RAM constantly consumes power while the device is on to refresh its contents – even if the device is performing no other functions. There are temporary storage options that use less power and keep their

future memory
IBM is one of the many companies that is researching MRAM memory technology, which will be ultrafast, less power-hungry, and retain stored data after a computer has been shut down.

information when turned off, but they are slower than RAM and generally have smaller capacity. A replacement for existing RAM technology called Magnetic RAM (MRAM) will almost certainly become available within a few years. MRAM stores more information, accesses it faster, consumes less power, and retains information when not connected to a power source. One of the implications is that computers will no longer need to be "booted up."

holographic memory and smart clothes

Hard disks used in today's digital devices are large and noisy, and they consume a lot of power. Those in expensive personal computers are typically able to store about 100 GB, and they transfer bits at a rate of a few tens of megabytes per second. An emerging technology, called holographic memory, will be able to store about 1,000 GB in a disk the size of a CD and transfer information at the rate of several gigabytes per second.

holographic memory
A holographic writer will work by splitting a laser beam into an object beam and a reference beam. The object beam picks up digital data as it passes through a spatial light modulator, in which binary data is represented by light and dark squares. Each beam takes a different route to a crystal disk on which the data is etched by the interference pattern produced where the two halves of the beam meet at an angle.

Modern CPUs are much faster than their equivalent of ten years ago – but there is still room for improvement. Integrated circuit technology is progressing in two main directions. First, the number of transistors etched onto integrated circuits is increasing, thanks to advances in miniaturization. Second, circuits can be built into different materials. For example, a new process called desktop fabrication will produce low-cost, integrated circuits on a special plastic material. It will be possible to use the resulting processors in a wide range of devices, perhaps integrating them into wearable technology, or "smart clothing." In general, the processors of the future will be much more powerful, yet use a lot less electrical power. This will make them quieter as they will no longer need a noisy fan to keep them cool.

wearable computers
Clothing, known as smart clothing, is becoming available, and will have built-in mobile phones, digital music players, or even personal computers.

electronic paper

Electronic paper is a display device consisting of a thin plastic sheet impregnated with spherical globules, which are half-white and half-colored. A digital signal rotates them, so that they show their white or colored side. These can display any text or image, and power is needed only when the display is changed. This technology, also known as electronic ink, is already in use as microcapsules containing black and white particles that are alternated by an electrical charge. This technology changes too slowly to display digital video. However, electronic paper will soon look and feel like ordinary paper, and will be used for large, high-resolution, low-power, and relatively cheap, wall-hanging digital displays.

lcd and plasma screens

The integration of digital networks and television has to overcome the problem that television displays are large, power-intensive analog devices, using a high-voltage electron gun called a cathode ray tube (CRT). There already exists a range of alternatives to CRT displays. For example, laptop computers use flat, liquid crystal display (LCD) screens, which are viewable from a wide angle, are bright, and consume much less power than CRT displays. But they are also much more expensive, and the cost increases rapidly with size. An alternative is the plasma display. This is similar to an LCD screen, but is better-suited to large displays, although costs remain prohibitive.

The resolution of a display is also important, and a sophisticated digital screen is able to display high-definition television (HDTV) pictures, composed of many more pixels than a conventional television picture.

new connections
Solar-powered aircraft may soon be circling above major cities, providing reliable, ultrafast, connection to the internet.

satellite networks

New developments in networking technology are also planned. First, a range of digital devices will be able to access broadband digital services via a network of low, earth-orbiting satellites. Today, most communications satellites occupy high, geostationary orbits and remain apparently still above any point over the equator. This enables anyone with a communications dish to exchange digitally encoded information with them. But because geostationary satellites are so far above the earth's surface, signals must be high powered, making it impossible to access network services without a dish. With a network of satellites in much lower orbits, ordinary handheld devices will be able to access high-speed, wireless digital networks from any point on the planet. Second, solar-powered airships or airplanes are projected; these will circle large cities and provide a similar service.

bluetooth

Another major development in network technology is the ability to make any wireless connections between any digital devices. Bluetooth technology makes this possible: a Bluetooth-enabled network in a car can connect digital music players, mobile telephones, and handheld computers wirelessly to each other and the rest of the world. In the home, digital devices will become entirely portable, connecting whenever they're required.

smart appliances

Finally, connectivity will extend beyond the range of devices that we currently use. For example, refrigerators and microwave ovens that have permanent connections to the internet are already available, making it possible to do your shopping, or send and receive emails, from your kitchen. These "smart appliances" may seem extravagant and unnecessary to us at present, but in the more distant future, when the cost of digital devices is much lower, they will probably become part of our everyday lives, monitoring and attending to our needs.

remember to buy the milk...

the end of shopping
More and more devices will become connected – perhaps, one day, your fridge will be connected to the internet, and you will use it to order food as well as store it.

future trends

People make many predictions about the impact that digital technology will have on our future lifestyles. Some forecast the development of "immersive environments," which will deliver three-dimensional virtual reality and even odors across the internet. Others predict the emergence of a new generation of computing devices, called quantum computers, which will be able to process information at speeds undreamed of today.

No one can really know what the future holds. There is uncertainty in the technology itself – new discoveries may completely change the direction of technological change. There is uncertainty in people's behavior – perhaps people will not even want an all-digital future. And there is uncertainty in the global economy. But in spite of the uncertain nature of our technological future, many experts have confidently set out their vision of how life may be different when we are surrounded by integrated digital technologies.

clever dog
Artificial intelligence has already found its way into high-tech toys. In 2000, a Japanese company introduced a computerized dog that responds to voice commands and recognizes its name and its owners' faces.

digital intelligence

New approaches to computer programming and design have led to systems that show some intelligence. Computers are not inherently intelligent: they simply follow programs. But even today, some computers can learn to recognize faces or handwriting, while others help to run air-traffic control systems or autonomously explore other planets. Speech recognition is already a reality, although it still makes mistakes. In the future, however, the mistakes will be few and far between, and this will probably render keyboards redundant. It will also enable more flexible, humanlike interaction with digital devices – for example, your home network may search the internet and suggest information that it thinks might be useful, without having to be asked. Perhaps computers will be able to speak with voices that sound human, and will be aware of subtle movements of a person's body or eyes and be able to respond to them. In short, as computers become more powerful, so machine – or artificial – intelligence may become very much more like human intelligence.

the price of progress

Most experts agree that implementing digital technology on a global scale will result in the cost of information and services falling dramatically – potentially bringing the benefits of digital technology to many more people on the planet. At present, most people would agree we are in a transition period. The development of each new technology requires massive initial funding, which can only be recouped by high prices for the first few years. By the time the costs of the new technology have fallen far enough for more people to have access to it, the technology has become outdated or obsolete. This built-in obsolescence is inevitable while the pace of technological advance is so great, and it is perhaps an economic necessity in the onward march of digital techno-logy, with all the benefits and new possibilities that it could bring.

writing in
The computer keyboard is likely to become increasingly obsolete, as voice- and handwriting-recognition improves and becomes more widespread.

digital evolution
Nearly all digital devices have been superseded by more powerful versions. This is frustrating for users, and causes concern among environ-mentalists, but seems to be an inevitable consequence of the speed of the digital revolution.

glossary

3G
Acronym for "third generation" – a collection of digital services, which includes digital video and is accessible via suitably equipped mobile devices, such as mobile phones.

analog
Any representation of text, sound, or images that does not involve digitization.

ADSL
Acronym for "Asymmetric Digital Subscriber Line." See DSL (Digital Subscriber Line).

ASCII
Acronym for "American Standard Code for Information Interchange" – a code in which each character (either a number, letter, keyed character, or command) is represented as a number that is translated into binary code and is used by computers and printers.

binary
A system of representing numbers that has 2 as its base and uses only two digits: 0 and 1. Computers use binary code as it works well with digital electronics and Boolean algebra.

bit
Abbreviation of "binary digit" – the smallest unit of information a computer can hold, and has the value of either 1 or 0.

bitmap
The main type of digital image where a grid of squares is placed over an image and each square is assigned a number determined by the brightness of the square.

Bluetooth
A system of connecting digital devices without cables by using high-frequency radio waves carrying digital signals.

broadband
Describes any connection between digital devices, such as internet access via DSL or cable modem, in which a large amount of digital information is transferred every second.

byte
A group of eight bits. There are 256 different ways to combine eight bits, so there are 256 possible bytes.

CCD
Acronym for "Charge Coupled Device" – an electronic memory in which semiconductors can be charged by light or electricity. One use of charge coupled devices is to store images in digital cameras, video cameras, and optical scanners.

codec
Abbreviation of "coder/decoder" – a device that converts analog signals to digital to be read by a computer and converts the digital signals back to analog.

compression
Any of a variety of techniques of reducing the amount of digital information needed to represent text, sound, images and video. See MP3.

CPU
Acronym for "Central Processing Unit" – the CPU controls a computer and contains units that perform arithmetical and logical operations, and interpret and execute instructions.

download
The transfer of digital data across a network, typically a computer file via the internet.

DSL
Acronym for "Digital Subscriber Line" – a method of broadband connection to the internet.

DVD
Acronym for "Digital Versatile Disc" – a method of storing digital information, which looks like a CD, but which has a far greater storage capacity.

email
Abbreviation of "electronic mail" – text documents, often with other digital information attached, sent between individuals on a network, most usually across the internet.

ethernet
The most common network protocol. Ethernet networks are the most common type of local area network.

handshaking
The exchange of information between digital devices when they first connect to form a network.

holographic memory
A new development in high-capacity digital storage, in which digital information is encoded as three-dimensional patterns in a crystalline disc.

HTML
Acronym for "Hypertext Markup Language" – the computer language used to write pages of information on the World Wide Web.

hub
A central device that connects several computers or networks together. A hub may simply pass on data, and is known as "passive," or it may be "active" and amplify the data for long-distance connections.

hyperlink
A link between documents written in HTML that are accessible across a network, usually the internet.

internet
The definitive digital network, connecting many millions of computers located on every continent.

IP address
A number used to identify any computer connected to the internet.

ISP
Acronym for "Internet Service Provider" – a company providing internet access for organizations and for individuals.

LAN
Acronym for "Local Area Network" – a relatively small number of connected computers and other digital devices, typically within the same building.

modem
Abbreviation of "modulator-demodulator" – a device that encodes digital information so that it can be transferred via telephone

MP3
Abbreviation of "MPEG Layer 3" – a compressed digital sound format that is ideal for downloading music from the internet.

MPEG
Acronym for "Moving Picture Expert Group" – an organization that develops effective ways of compressing digital information to reduce the amount of information that is needed to represent sound or video with the minimum loss of quality.

MRAM
Acronym for "Magnetic Random Access Memory" – a recently developed technology that uses less power and has more storage capacity than traditional RAM, and which is based on integrated circuits.

network
A group of interconnected digital devices.

PDA
Acronym for "Personal Digital Assistant" – a device, typically hand-held, which acts as an electronic organizer, handheld computer, and a mobile communications device.

pixel
Any of the tiny picture elements that make up a bitmapped, digital representation of an image.

protocol
A specification of the rules and format in which digital information is to be transferred across a computer network.

RAM
Acronym for "Random Access Memory" – the working memory of a computer. RAM is used for storing data temporarily while the data is in use and for running software applications. The data held in RAM is lost when the power is switched off.

router
A router receives an electronic message over a

network, notes the traffic load and the number of hops to the destination of the message, and then determines the best possible path for the message to reach that destination.

sampling
Taking the value of a signal at evenly spaced intervals in time – it is the first of three steps in the process of digitizing an analog signal. The other two steps are quantizing the signal and then encoding it.

SCSI
Acronym for "Small Computer Systems Interface" – a high-speed connection to devices such as hard drives, CD-ROM drives, floppy drives, scanners, and printers.

semiconductor
Materials including silicon and germanium that are poor conductors of electricity, unlike copper and are poor insulators, unlike rubber. Computer chips are made of semiconductor materials. Semiconductors make it possible to miniaturize electronic components, such as transistors. Miniaturization means that components are smaller, faster, and use less energy.

server
A computer that holds digital information shared between users connected to a network.

streaming
The transfer of live or recorded sound or video, which allows users of a network – typically the internet – to listen to digital sound or watch digitized video without first having to download it in its entirety.

TCP/IP
Acronym for "Transfer Control Protocol/Internet Protocol" – a set of agreed rules by which digital information is transferred across the internet.

transistor
An electronic switch in computer circuitry, which is either on or off depending on whether current is flowing between two terminals known as the source and the drain. The flow of current is controlled by an electric field, which varies depending on whether a positive or negative charge is sent to a third terminal in the transistor called the gate.

URL
Acronym for "Uniform Resource Locator" – an address for a particular HTML document, sound, or other resource available on the internet. URLs typically begin with "http."

USB
Acronym for "Universal Serial Bus" – which can connect up to 127 devices using inexpensive cable that can be up to 16 ft (5m) long.

virus
A small, malicious, computer program that transfers itself between users of a network, typically via email, with the intention of causing harm to users' computers. All viruses are created intentionally.

WAP
Acronym for "Wireless Application Protocol" – the protocol used to connect mobile devices to the internet.

web page
A text document, written in HTML, which is decoded and viewed via a computer program called a browser. A web page can contain text, images, sound, and video, and perhaps most importantly, links to other web pages.

website
A collection of interlinked web pages, all published on the same server.

World Wide Web
Alternatively, "the Web" – a virtual, ever-changing collection of interlinked information made available to anyone connected to the internet. Any piece of information on the Web has a URL to identify it.

worm
Similar to a virus, a worm is a computer program that copies itself to many computers on a network, typically the internet, causing them to slow down.

index

Further reading

The Rough Guide to the Internet, Angus J Kennedy, Rough Guides, 2001. ISBN: 1858285518

Where Wizards Stay up Late, Katie Heffner and Matthew Lyon, Simon & Schuster Inc., 1998. ISBN: 0684832674

MP3: The Definitive Guide, Scott Hacker, O'Reilly UK, 2000. ISBN: 1565926617

The Official MP3.com Guide to MP3, Michael Robertson and Ron Simpson, MP3.com, 1999. ISBN: 096705740X

Weaving the Web, Tim Berners-Lee, Texere Publishing, 2000. ISBN: 1587990180

A Brief History of the Future, John Naughton, Phoenix Press, 2000. ISBN: 075381093X

The Road Ahead, Bill Gates, Penguin Books, 1996 (revised edition). ISBN: 0140243518

Being Digital, Nicholas Negroponte, Coronet, 1996. ISBN: 0340649305

Parent's Guide to Protecting Children in Cyberspace, Parry Aftab, McGraw-Hill Publishing Company, 2000. ISBN: 0077096746

See also the Dorling Kindersley *Essential Computers* series, which includes practical information relating to many of the topics covered in this book.

Selected online resources:
http://www.cochran.com/ start/guide/
http://www.cs.berkeley.edu/ ~russell/ai.html
http://www.design.philips. com/vof/toc1/home.htm

....

Author's acknowledgments
I would like to thank Miki Lindley for supporting me while I wrote this book. Also, a vote of appreciation goes to Ian Whitelaw, Simon Avery, and John Watson at Design Revolution, and to Peter Frances at Dorling Kindersley.

Illustration
Richard Tibbitts and Martin Woodward, AntBits illustration

Index
Indexing Specialists, Hove

Additional picture research
Penni Bickle

Picture credits
Central Museum Utrecht: 22. **Eink Corporation**: 61(b). **IBM**: 59. **Image Bank**: Front cover, 1. **Natural History Museum**: 23(b). **Nokia Corporation**: 56. **Paul Mattock of APM, Brighton**: 15(tc), 19, 20(tl), 30(bl), 36(tl/cr). **Photonica**: Kaz Chiba 31(tr), 51(br). **Science Museum**: 39(tr). **Science Photo Library**: Rosefeld Images LTD 4, 38(bl); Mike Bluestone 5, 8; George Bernard 11, 36(br); Hank Morgan 45; Sam Ogden 55; jerrican 58; Lawrence Livermore National Laboratory 63; Peter Menzel 64. **Simon Avery & Kree**: 6, 15(br), 18(bl), 26, 32(cl), 47(tr), 53, 54, 65(tr, br). **Telegraph Colour Library:** 57

Dorling Kindersley would like to thank Microsoft Corporation for permission to reproduce screen images from within Microsoft® Windows® Millennium Edition.